North American Dinosaurs
TYRANNOSAURUS REX

Anastasia Suen

rourkeeducationalmedia.com

About The Author

A former fifth grade teacher, Anastasia Suen is the author of 78 books for young readers. She teaches writing at Southern Methodist University in Texas.

© 2007 Rourke Educational Media

All rights reserved. No part of this book may be reproduced or utilized in any form or by any means, electronic or mechanical including photocopying, recording, or by any information storage and retrieval system without permission in writing from the publisher.

www.rourkeeducationalmedia.com

Photos/Illustrations: Cover © Joe Tucciarone; title page © The Field Museum (GEO86284C); pages 4, 13 © Jan Sovak; pages 6, 14 © Train Hall Studios; pages 9, 21 © Francois Gohier; page 10 © Christopher P Srnka; page 17 © Julius Csotonyl; page 18 © Gerhard Goeggemann; page 22 © American Museum of Natural History Library

Editor: Robert Stengard-Olliges

Cover and page design by Nicola Stratford

Library of Congress Cataloging-in-Publication Data

Suen, Anastasia.
 Tyrannosaurus rex / by Anastasia Suen.
  p. cm. -- (North American dinosaurs)
 ISBN 978-1-60044-255-1 (hardcover)
 ISBN 978-1-60472-994-8 (paperback)
 ISBN 978-1-60472-064-8 (Lap Book)
 ISBN 978-1-60472-121-8 (eBook)
1. Tyrannosaurus rex--Juvenile literature. I. Title. II. Series.

QE862.S3S84 2007
567.912'9--dc22

2006016339

Rourke Educational Media
Printed in the United States of America,
North Mankato, Minnesota

rourkeeducationalmedia.com
customerservice@rourkeeducationalmedia.com • PO Box 643328 Vero Beach, Florida 32964

Table of Contents

Tyrant Lizard King	4
A Healthy Appetite	6
A Big Head	8
Chomping and Chewing	11
Reaching and Running	12
Heads or Tails?	18
Lost!	20
Found!	22
Glossary	23
Further Reading/Website	24
Index	24

Tyrant Lizard King

Tyrannosaurus rex was the king of the dinosaurs. It was the largest meat-eating dinosaur on Earth. This dinosaur was as long as an 18-wheel semi truck. It was as tall as three men.

Its name tells you the story. *Rex* means king. But this dinosaur wasn't a kind ruler. *Tyrannosaurus* means "tyrant lizard." A **tyrant** is a cruel ruler.

Tyrannosaurus rex *was the largest meat-eating dinosaur.*

Why was *Tyrannosaurus rex* cruel? It ate other dinosaurs!

But don't worry. This dinosaur won't eat you. It lived more than 65 million years ago. All that's left now are the bones.

Tyrannosaurus rex bones have been found in the western United States and Canada. In some places there are lots of bones—almost the entire dinosaur is there. In other places, only a few bones are left.

Tyrannosaurus rex skin marks have been found on rocks. The rocks had rows of raised bumps. But the rocks don't tell us what color this dinosaur was. We can only guess.

A Healthy Appetite

Tyrannosaurus rex, or *T. rex*, was a meat eater. It ate other dinosaurs. Animals that eat meat are called **carnivores**. Dinosaurs that ate meat are called **carnosaurs**.

T. rex wasn't the only dinosaur that ate meat. It wasn't the only dinosaur that ate other dinosaurs. But it was the largest. In fact, *T. rex* was the largest carnosaur on Earth.

A typical *T. rex* weighed more than five tons (4,536 kg). In order for this big animal to survive, it had to eat a lot of food. *T. rex* ate more than five tons of meat a week. It ate as much as it weighed!

Each week Tyrannosaurus rex *ate its weight in meat.*

A Big Head

T. rex had a very large head. It was more than four feet (1.2 m) long. That's the size of a baby's crib!

A big head is very heavy. The *T. rex* skull at Chicago's Field Museum weighs 600 pounds (272 kg). It's so heavy that the museum keeps the skull on a stand by itself. The skull was too heavy for the steel holding up the rest of the skeleton. A cast of the head is displayed on the skeleton.

Inside this huge head was a tiny brain. The space inside the skull where the brain would have fit is only about the size of a quart of milk.

This Tyrannosaurus rex *skull shows how large the dinosaur's head was.*

Chomping and Chewing

When *T. rex* opened its big mouth, watch out! *Tyrannosaurus rex* had fifty to sixty teeth. These teeth were the size of a banana, about six to 12 inches (15 to 30 cm) long. But the teeth weren't soft like a banana. *T. rex's* teeth were like steak knives. They had jagged edges that could cut through skin and bone.

Eating five tons (4,536 pounds) of meat each week wore down these teeth. But unlike humans, who have only baby teeth and then adult teeth, *T. rex* had new teeth coming in all the time. It was always ready to eat!

Tyrannosaurus rex *had new teeth growing in all the time.*

Reaching and Running

For such a big animal, *Tyrannosaurus rex* had two very short arms. Each arm had two fingers with sharp claws, but *T. rex* couldn't use them for eating. Its arms were too short to reach its mouth.

Try this:
1. Put your arms out to the side.
2. Now have someone measure from fingertip to fingertip.

Your arm span is almost the same as your height! This wasn't true for T. rex.

A *Tyrannasaurus rex* could be 50 feet (15 m) long from head to toe. Its legs alone were 13 to 20 feet (four to six m) high. So why were its arms only three feet (one m) long? No one knows.

Scientists aren't sure how T. rex used its short front arms.

T. rex *probably had to chase and catch its food.*

Scientists think they know more about how *Tyrannosaurus rex* used its two back legs. *T. rex* was a **theropod**. It was a meat-eating dinosaur that walked on two legs. These two long legs helped it chase its prey. *T. rex* was a **predator**. It had to catch its food before eating it.

Or did *T. rex* just take other dinosaur's dinners away? Did another animal make the kill only to have *Tyrannosaurus rex* take the food away because it was bigger? Scientists can't decide. It's hard to say more than 65 millions years later.

> **Talons**
> T. rex had feet like a bird. Three toes pointed forward. One toe pointed back. The three forward toes had talons. These claws were eight inches long.
>
> Just like a bird, T. rex used its feet and its mouth to attack. The talons on its feet were very sharp!

Tyrannosaurus rex *had talons, or claws, on its feet.*

Heads or Tails

T. rex had a long tail. The tail on the *Tyrannosaurus rex* skeleton at the Field Museum is twenty feet long!

Now scientists see it a new way. They think that *T. rex* probably ran leaning forward. Its head was down, so its tail was up. Holding up its tail helped *T. rex* keep its balance. It also helped *T. rex* run

Today, museums are rebuilding their *T. rex* skeletons. They are moving the head down and the tail up.

For many years, scientists thought that *T. rex* dragged its tail on the ground. They thought the tail had to be down so *T. rex* could hold up its heavy head.

Lost!

Why aren't there any *T. rexes* here today? Why are they **extinct**? Some scientists think it may be because of an **asteroid**.

One possible explanation is that 65 million years ago, a giant asteroid crashed into Earth. The crash made a huge cloud of dust. The dust blocked out the sun. This made the weather colder.

Before the asteroid, Earth was very warm. After the asteroid crashed, the weather turned colder. Lots of plants and animals died. All that was left of the animals were the bones. Soon these bones were covered with dirt. Now, after millions of years, the only way to find the bones is to dig for them.

Digging Up Bones

Dinosaur hunters have to dig up the bones they find. First the fossil hunters take away all of the extra dirt around the bones. Then they brush the bones carefully. When the bones are clean, the scientists cover them with plaster. The plaster coating protects the bones. The hunters take the bones to a museum. There they take off the plaster and put the bones together.

Found!

Barnum Brown was a dinosaur hunter. Newspaper reporters called him "Mr. Bones." He discovered the first *Tyrannosaurus rex* bones in 1902 in a creek in Montana.

Workers put the bones on a wagon. Horses pulled the wagon to the railroad line. A train took the bones all the way to New York City. In 1908, Brown went back to the creek. He found more *T. rex* bones.

At the museum, workers put the bones together to show how *Tyrannosaurus rex* would have looked when it was alive. In 1915, this T. rex skeleton was ready. You can still see it in New York at the American Museum of Natural History.

This skeleton shows how **Tyrannosaurus rex** *might have looked when it was alive.*

Glossary

asteroid (AS teh roid) — a rock in space

carnivore (KAR nuh vor) — an animal that eats meat

carnosaur (KAR nuh sor) — a meat-eating dinosaur with two legs

extinct (EK stingkt) — a type of animal or plant that has died out

predator (PRED uh tur) — an animal that lives by killing and eating other animals

theropod (THIR uh pahd) — a meat-eating dinosaur that walked on two legs

tyrant (TYE ruhnt) — a cruel leader

Tyrannosaurus fossils have been found in the following sedimentary rocks:
- the *Lance Formation* of Eastern Wyoming;
- the *Hell Creek Formation* of Eastern Montana, Southwestern North Dakota, and Northwestern South Dakota;
- the *Livingston Formation* of Montana;
- the *Javelina Formation* of Big Bend Texas;
- the *Laramie Formation* of Colorado;
- the *McRae Formation* of New Mexico;
- the *Scollard* and *Willow Creek formations* of Alberta, Canada;
- the *Frenchman Formation* of Saskatchewan, Canada.

Written by: Dr. Philip Bjork, Museum of Geology, South Dakota School of Mines and Technology, Rapid City, SD 57701. 1995.
Map reference: http://www.northern.edu/natsource/earth/T. rex1.htm

Index

American Museum of
  Natural History   22
arms   12
brain   8
Brown, Barnum   22
carnivore   6
carnosaur   6
Chicago Field Museum
  8, 18
plaster   21
predator   15
skull   8
tail   18
talons   16
teeth   11
theropod   15

Further Reading

Dahl, Michael. *T. rex: The Adventure of Tyrannosaurus rex.* Picture Window Books, 2004.

Schomp, Virginia. *Tyrannosaurus and Other Giant Meat-Eaters.* Benchmark Books, 2003.

Skrepnik, Michael. *Tyrannosaurus-rex—fierce King of the Dinosaur* Enslow Publishers, 2005.

Websites to Visit

www.amnh.org/exhibitions/expeditions/treasure_fossil/Treasures/Tyrannosaurus/tyrannos.html?dinos

www.royalsaskmuseum.ca/about/museum_history_scotty.shtml

www.fieldmuseum.org/sue/